Plan Bee

The #1 Guide to Beekeeping, Apiary Husbandry, & Colony Collapse Management

by

Austin Robinson, BBL
Connor Gleim, Bee-hD
Patrick Golden, ADHD

www.potholepress.com

Book design by Rebecca

First published on May 20th, 2023

ISBN:
978-0-9992029-8-2

Some names and characters of people, places, and things have been wildly altered because this is – for the most part – a satirical novel that holds the sole purpose of determining if we can become #1 Bestselling Authors in an underutilized sub-sub-sub-category of a major online marketplace.

THIS BOOK IS DEDICATED
TO OUR LOVE: **BEES**

NOTE: ALL PROCEEDS FROM THE SALES OF THIS BOOK WILL BE DONATED STRAIGHT TO THE OREGON MASTER BEEKEEPER PROGRAM, A COOPERATIVE PROGRAM BETWEEN OREGON STATE UNIVERSITY AND THE OREGON STATE BEEKEEPERS ASSOCIATION BASED IN THE UNITED STATES DEDICATED TO BRINGING PEOPLE CLOSER TO BEES THROUGH EDUCATION AND AWARENESS, INSPIRING EMPATHY AND ACTION FOR APIARY HEALTH AND THE WELLBEING OF POLLINATORS.

LEARN MORE ABOUT THE OREGON MASTER BEEKEEPER PROGRAM HERE: EXTENSION.OREGONSTATE.EDU/MB

<u>TABLE OF CONTENTS</u>

ACKNOWLEDGMENTS

We would like to acknowledge Marshall Dahng Geyer for giving us the hope and willpower and money to complete this novel. So much money. Like, a bee of honey. We love you, Marshall. Thank you.

INTRODUCTION 1
THE BUZZ ABOUT US

"There's getting buzzed,
and then there's getting BUZZED!"
- Barry. Yes, that one.

NOTE: The sole point of this chapter is to recount the authors' experiences with bees. If you would like to learn more about Apiary Husbandry, it is best to skip this chapter. Although, who knows, you may just find inspiration in our stories and get into Beekeeping if you haven't already!

Connor

After my 17th divorce, I decided to see what all the buzz is about. About bees, that is. I had heard about bees before, but did I really know about bees? They seem to be loyal companions to humankind – always there, making noises and spitting on our flowers. Sure, there are times they become temperamental and back their ass up into my arm, leaving behind their stinger that I didn't ask for… but then they did make sure to die as an apology. Not like Addison – my ex-wife – who probably would have just continued living after stabbing me with a sharp object… again. No, bees have dignity. I know deep down bees are always the wife I never had. My 18th wife. That was, of course, until I met Stella – my real 18th wife.

I was 40 years old when I decided to attend the best apiary institution in the world: The Oregon Master Beekeeper Program in Covallis, Oregon! After a long series of failed career attempts – political comptroller, home masonry repair man, ghostwriter; just to name a few – I wanted to become a researcher and master of the bee! I have been buzzing since I was just a little baby. My mother even claims that I came out of her womb in a bumblebee outfit screaming "BUZZ BUZZ BUZZ!" while fighting off other pollinators. This, of course, is ridiculous to imagine, but if you know me, you know that it is much more realistic than how a baby naturally comes out!

During my first semester in the Beekeeper Program, I met this wonderful woman named Deptember – Dep for short. She sat next to me in Let's Get Buzzed, a class taught by Professor Bristina Hivenhaus. The course was an intro-level one to mine and Dep's 'Don't Pollute, Pollinate!' track of our graduate degree. The moment I fell in love with her was when she leaned over to me and said, "So honey isn't just bee cum, huh? Can you bee-lieve this!"

From there, it was as smooth as spreading honey on a bagel. I received a 4.0 GPA, graduated at the top of my class right under Dep, worked with the United Nations to secure better Apriary Law codes in the Pacific Northwest of the United States, and fell in love with bees even more. At the age of 45, I will be presenting my research regarding Apiary Husbandry & Colony Collapse Disorder Management to the United Nations on World Bee Day! And who knows, there may even be mini Connor and Deptember Gleims on the way ;)

Austin

Coquille Valley. That's where it all began. My first kiss, my first sip of alcohol, my first touch of a bee. When I was just six years old, my father – a single parent – would hike along the banks of the Coquille River every day searching for my mother, who was trampled by a large family of beavers (both in numbers and size) rushing away from a beehive they had accidentally bumped into. And I was right beside him.

My mother gave birth to me on the river while making her way back from Bandon, Oregon. My parents had an affinity for the Coos Tribe and spent their honeymoon on their tribal land after performing a shotgun wedding in the Coosan language by accident. She tried to hold off giving birth to me until they reached inland, and she was almost successful. Until an elderly couple on a river steamboat noticed that her water broke while she was climbing a really, really big rock – I'm sure my mom just assumed her coochie got wet from the river water. Suddenly, she was giving birth to a beautiful baby boy: Me.

Anyway, I stood by my father's side as we did that hike single day for five years: from my ages of six to eleven. He really thought she was still out there. Waiting.

I didn't start school until I was 12 years old. I was much older than the other Kindergarteners. By the time I was in Grade 1, I was a teenager. I turned 18 the day I started Grade 6. You get where I'm going with this: I was out of place for most of my life. Graduating secondary school at the age of 24 is not something you easily get over. Sure, my classmates were jealous of my facial hair and chest hair in Grade 4, I easily became the national champion of the Oregon Beekeeping League for five years in a row, and I was legally allowed to have sex with my teachers in high school. But there were many downfalls, too. Sarah – my girlfriend in 3rd grade – broke up with me after her parents got wind of how old I was (15), and maintaining a fast food job without having learned multiplication yet was pretty tough. But I made it through. And I kept my eye on the prize: using bees to sting the hell out of beavers.

Once I knew I wanted to make an entire animal species suffer from bee stings, I had my college plans. I went to the University of Montana in Missoula. Throughout my entire Beekeeping program, my professors and peers told me I was stupid and that I couldn't train thousands, if not millions, of bees to specifically target the beaver population. But I knew better. Now look at me! I'm a Master Beekeeper who is currently housing 1,500 beehives in my backyard illegally!

Patrick

I just love honey.

INTRODUCTION 2
BEES IN THE TRAP

"Bee this, bee that. How about you BEE a good person and leave bees alone!"

- Oprah. No, not that one.

Bees are the prime essential pollinators for our planet. They ensure the production of many fruits, vegetables, and nuts that we rely on for food. In fact, one-third of the food we eat is pollinated by bees.

Bees are social insects that live in colonies. A colony consists of the queen bee, a few hundred drones (male bees), and thousands of worker bees. Worker bees are responsible for all tasks necessary to keep the colony alive, including gathering food, building the hive, and caring for the queen and the young bees.

<u>Bees are facing a number of threats, including:</u>

- **Climate Change**: Climate change is causing the environment to become more difficult for bees to survive. For example, warmer temperatures cause bees to become more active earlier in the spring, which can lead to them running out of food before the end of the season.

- **Habitat Loss**: Habitat loss is another major threat to bees. As we continue to develop land, we are destroying the natural habitats that bees need to survive, which includes wildflower meadows that provide bees with food and shelter.

- **Pesticides**: Pesticides are a major threat to bees. Many pesticides are toxic to bees, and even those that are not toxic can weaken bees and make them more susceptible to disease.

- **Varroa Mites**: Varroa mites are tiny parasites that feed on the blood of bees. They can weaken bees and make them more susceptible to disease.

- **Nosema Ceranae**: Nosema ceranae is a fungus that can infect bees and cause them to become sick.

- **Colony Collapse Disorder**: CCD is a condition in which worker bees in a hive suddenly disappear. The cause of CCD is not fully understood, but it is thought to be a combination of factors, including the ones mentioned above.

The decline of bees is a serious problem. If the bee population continues to decline, it will have a significant impact on our food supply. We may see a decrease in the availability of fruits, vegetables, and nuts, and the prices of these foods may go up.

There are things we can do to help bees. We can support beekeepers, plant bee-friendly flowers in our gardens, and reduce our use of pesticides. We can also advocate for policies that protect bees and their habitats.

Here are some additional things you can do:

- **Buy local honey**. When you buy local honey, you are supporting beekeepers who are helping to keep bees healthy and productive.

- **Avoid using pesticides** in your garden. If you must use pesticides, choose those that are labeled "bee-friendly."

- **Plant bee-friendly flowers** in your garden. Bees love flowers that are native to your area. You can find a list of bee-friendly flowers online or at your local nursery.

- **Educate others** about the importance of bees. Talk to your friends, family, and neighbors about the threats facing bees and what they can do to help.

- **Support organizations** that are working to protect bees. There are many organizations working to protect bees and their habitats. You can support these organizations by donating money, volunteering your time, and/or spreading the word about their work.

This book provides a comprehensive guide to beekeeping, apiary husbandry, and colony collapse management. It covers topics such as:

1. The biology of honeybees
2. The types of beehives, how to manage them, & choosing a location
3. The equipment you need for beekeeping
4. Beekeeping safety, laws, & regulations
5. Harvesting honey
6. How to prevent and manage CCD
7. The Future of Beekeeping

1
BEE-OLOGY

"Fools liken honey to bee cum.
Experts know the difference."
- Very old Scottish Proverb

Honeybees are social insects that live in colonies. A colony consists of a queen bee, worker bees, and drones. The queen bee is the only fertile female in the colony. She lays the eggs that will become new bees. Worker bees are female bees that do not reproduce. They are responsible for all of the work in the colony, including gathering food, caring for the young, and building the hive. Drones are male bees that do not have stingers. Their only function is to mate with the queen bee.

Honeybees are found all over the world and there are over 40 different species. However, the most common species is the western honeybee (Apis mellifera). Western honeybees are native to Europe and Asia, but they have been introduced to other parts of the world, including North America, South America, Africa, and Australia.

Honeybees are important pollinators and play a vital role in the ecosystem. They help pollinate many different types of plants, including fruits, vegetables, and flowers. Without honeybees, many of these plants would not be able to reproduce and our food supply would be severely limited.

Honeybees have a number of specialized features that help them survive and thrive. They have a long, thin proboscis they use to suck up nectar from flowers. They also have a pair of compound eyes that allow them to see a wide range of wavelengths, including ultraviolet light. This helps them find flowers and assess the quality of the nectar.

Honeybees have a sophisticated communication system that allows them to share information with each other. They use a dance language to communicate the location of food sources. This dance language is very complex and it is still not fully understood.

Honeybees undergo complete metamorphosis, which means they go through four stages of development: egg, larva, pupa, and adult. The egg hatches into a larva, which is a worm-like creature that feeds on royal jelly or pollen and honey. After about four weeks, the larva pupates and enters the resting stage. The pupa transforms into an adult bee over the course of about two weeks.

Honeybees live in highly social colonies that can range in size from a few hundred to over 50,000 bees. The colony is organized into a hierarchy with the queen at the top. The queen is the only fertile female in the colony and is responsible for laying eggs. The workers are sterile females and perform all of the other tasks in the colony; such as gathering food, building the nest, and caring for the young. Drones are males and their only function is to mate with the queen.

Honeybees communicate with each other using a variety of signals, including pheromones, dances, and sounds.

Pheromones are chemical signals that bees use to communicate about food sources, the location of the nest, and other important information. Dances are a complex form of communication that bees use to share information about the location of food sources. Sounds are also used by bees to communicate, but they are not as well understood as pheromones and dances.

Honeybees are important for a number of reasons. They are important pollinators and play a vital role in the ecosystem. They also produce honey, which is a valuable food source for humans. Honeybees are also used in research and they have been used to develop new drugs and treatments for diseases.

Honeybees are fascinating creatures that play an important role in the world. They are important pollinators, they produce honey, and they are used in research. Honeybees are facing a number of challenges, such as habitat loss and climate change. However, there are a number of things that can be done to help honeybees, such as planting bee-friendly flowers and supporting beekeepers.

2
LOCATING, MANAGING, & SELECTING BEEHIVES

"Why HATE when you can polliNATE?"

- Did a bee write this?

There are many different types of beehives available. The most common type is the Langstroth hive. Langstroth hives are easy to manage and can be stacked on top of each other. Other types of beehives include top-bar hives, Warre hives, and skeps.

The Langstroth hive is the most popular type of beehive in the world. It is a vertical hive made up of a series of boxes that can be stacked on top of each other. The boxes are separated by frames, which are made of wood and have a foundation of wax or plastic. The bees build their honeycomb on the frames, which makes it easy for the beekeeper to harvest the honey.

The Langstroth hive is a versatile hive that can be used for both small and large beekeeping operations. It is also a relatively easy hive to manage, which makes it a good choice for beginners.

The top bar hive is a horizontal hive that is made up of a series of bars that are spaced evenly apart. The bees build their honeycomb on the bars, which makes it easy for the beekeeper to harvest the honey.

The top bar hive is a more natural hive than the Langstroth hive, and it is often preferred by beekeepers who want to give their bees more freedom. However, the top bar hive can be more difficult to manage than the Langstroth hive, and it is not as well-suited for large beekeeping operations.

The Warre hive is a horizontal hive that is similar to the top bar hive. However, the Warre hive uses shallower frames than the top bar hive, and the frames are not attached to the hive. This gives the bees more freedom to build their honeycomb, and also makes it easier for the beekeeper to harvest the honey. The Warre hive is a good choice for beekeepers who want to give their bees more freedom. However, it is not as well-suited for large beekeeping operations as the Langstroth hive.

There are many other types of beehives available, including the Kefir hive, the Flow hive, and the Warré-Langstroth hive. Each of these hives has its own unique advantages and disadvantages, and the best hive for you will depend on your individual needs and preferences.

When choosing a beehive, there are several factors to consider, including the size of your beekeeping operation, the type of honey you want to produce, and your personal preferences. If you are a beginner, it is a good idea to start with a Langstroth hive, as it is the most versatile and easy-to-manage hive. However, if you want to give your bees more freedom, you may want to consider a top bar hive or a Warre hive.

No matter what type of hive you choose, it is important to do your research and choose a hive that is well-suited for your needs. With a little bit of research, you can find the perfect hive for your bees and your beekeeping operation. We hope this book helps!

To reiterate: there are many different types of beehives available, so it is important to choose one that is well-suited for your needs. Some factors to consider include the size of your beekeeping operation, the type of honey you want to produce, and your climate.

- **Langstroth Hive**: The Langstroth hive is the most popular type of beehive in the world. It is a vertical hive that is made up of a series of boxes that can be stacked on top of each other. The boxes are separated by frames, which are made of wood and have a thin layer of wax or plastic. The bees build their honeycomb on the frames, which makes it easy for the beekeeper to harvest the honey.

- **Top Bar Hive**: The top bar hive is a horizontal hive that is made up of a series of bars that are spaced evenly apart. The bees build their honeycomb on the bars, which makes it easy for the beekeeper to harvest the honey.

- **Warre Hive**: The Warre hive is a horizontal hive that is similar to the top bar hive. However, the Warre hive uses shallower frames than the top bar hive, and the frames are not attached to the hive. This gives the bees more freedom to build their honeycomb, and it also makes it easier for the beekeeper to harvest the honey.

Once you have selected a beehive, you need to learn how to manage it. This includes tasks such as:

- **Inspecting the hive**: You should inspect your hive regularly to check on the health of the bees and to see how much honey they have produced.

- **Adding frames**: As the bees grow, you will need to add frames to the hive. This will give them more space to build their honeycomb.

- **Harvesting honey**: You can harvest honey from your hive once the bees have filled the honeycomb.

- **Winterizing the hive**: In the winter, you will need to winterize your hive to protect the bees from the cold.

There are many resources available to help you learn about beekeeping. These include books, magazines, websites, and beekeeping clubs.

By following these tips, you can select and manage a beehive successfully. Here are some additional tips for managing a beehive:

- Place the hive in a sunny spot that is protected from the wind.

- Provide the bees with a water source.

- Plant flowers that will attract bees.

- Be patient and gentle when working with the bees.

- Get to know your bees and learn their behavior.

With a little bit of care and attention, you can enjoy the many benefits of beekeeping for years to come.

The location of your beehive is important for the health and productivity of your hive. There are a number of factors to consider when choosing a location, including:

- **Sunlight**: Bees need sunlight to produce honey. The hive should be located in an area that receives at least 6 hours of sunlight per day.
    - The sun's rays help to warm the hive, which is important for the bees' development.
    - The sun also helps to dry the hive, which helps to prevent the spread of diseases.

- However, too much sunlight can also be a problem. If the hive is located in direct sunlight, the bees may overheat.

- **Water**: Bees need water to drink and to cool their hive. The hive should be located near a water source, such as a pond, stream, or birdbath.
  - Bees will not travel far for water, so it is important to make sure that there is a water source close to the hive.
  - The water source should be clean and free of chemicals.
  - If there is no water source nearby, you will need to provide a water source for your bees.

- **Wind**: Bees don't like too much wind, so avoid placing your hive in an area that is windy.
  - Wind can blow the bees off course, making it difficult for them to find their way back to the hive.

- Wind can also cause the hive to cool down too quickly, which can stress the bees.
- However, a little bit of wind can be beneficial for the hive. Wind helps to circulate the air around the hive, which helps to keep the hive cool and prevent the spread of diseases.

- **Shade**: Bees also don't like too much heat, so if you live in a hot climate, you may want to provide some shade for your hive.
  - The shade will help to keep the hive cool and prevent the bees from overheating.
  - The shade can also help to protect the hive from the sun's rays, which can damage the hive's structure.
  - However, too much shade can also be a problem. If the hive is located in too much shade, the bees may not be able to get enough sunlight to produce honey.

- **Privacy**: Bees are not aggressive, but you may want to place your hive in a private location so that you don't have to worry about people or pets disturbing it.
    - If you live in a busy area, you may want to consider placing your hive in a backyard or garden.
    - If you live in a rural area, you may be able to place your hive in a field or meadow.
    - However, if you live in a very rural area, you may need to be more careful about placing your hive in a private location. Bears and other wild animals may be attracted to the hive.

- **Proximity to flowers**: Bees need flowers to collect nectar and pollen. The hive should be located in an area that has a variety of flowers that bloom throughout the year.
    - The flowers should be close enough to the hive that the bees can easily reach them.

  - The flowers should be a variety of colors and types, so that the bees have a variety of food sources.
  - If you live in an area that does not have a lot of flowers, you may need to plant some flowers near the hive.

- **Laws and regulations**: Check with your local government to make sure that there are no laws or regulations that restrict where you can keep bees. There is more information on this later on in this book!
  - Some cities and towns have restrictions on the number of hives that you can keep.
  - Some cities and towns also have restrictions on the location of hives, such as requiring them to be a certain distance from schools or hospitals.

<u>Once you have considered</u> all of these factors, you can start looking for a location for your hive. Here are a few tips for finding a good location:

- **Talk to other beekeepers**: They can give you advice on good locations and help you avoid any potential problems.

- **Look for areas with a variety of flowers**: This will give your bees a good source of nectar and pollen.

- **Avoid areas with a lot of traffic**: This will help to keep your bees safe from being disturbed.

- **Inspect the site for pests and diseases**: Make sure that the area is free of pests and diseases that could harm your bees.

3
B.E.E. BEEKEEPING EQUIPMENT ESSENTIALS

"I wish I was a bee so bad, please somebody take me out of my misery."

- Scott

Beekeeping is a rewarding hobby that can also be a source of income. However, before you can start beekeeping, you need to gather the necessary equipment. Here is a list of the essential equipment you will need for beekeeping:

- **Beehive**: The beehive is the home of your bees. There are many different types of beehives available, so you will need to choose one that is well-suited for your needs.

- **Frames**: The frames are the foundation on which the bees build their honeycomb. They are made of wood and have a thin layer of wax or plastic.

- **Foundation**: The foundation is the wax or plastic sheet that is attached to the frames. It provides a guide for the bees to build their honeycomb.

- **Smoker**: The smoker is used to calm the bees before you work on the hive. It creates a cloud of smoke that masks the bees' alarm pheromones.

- **Hive tool**: The hive tool is a small, flat tool that is used to pry open the hive and to remove and replace frames.

- **Veil and gloves**: Protective gear is essential for beekeeping. The veil will protect your face and neck from bee stings, and the gloves will protect your hands.

- **Bee suit**: A bee suit is a one-piece garment that provides full-body protection from bee stings. It is not essential for beekeeping, but it can be a good option for beginners or for beekeepers who are allergic to bee stings.

- **Other supplies**: There are a few other supplies that you may need for beekeeping, such as a honey extractor, a hive mat, and a hive stand.

In addition to the essential equipment, there are a few other items that you may want to consider purchasing, such as:

- **Beekeeping book**: There are many great books available that can teach you the basics of beekeeping.

- **Beekeeping magazine**: Beekeeping magazines are a great way to stay up-to-date on the latest beekeeping news and techniques.

- **Beekeeping club**: Joining a beekeeping club is a great way to learn from other beekeepers and to get help and advice.

With the right equipment and a little bit of knowledge, you can start beekeeping and enjoy the many benefits that this hobby has to offer.

Here are some additional tips for choosing and using beekeeping equipment:

- When choosing a beehive, it is important to consider the size of your beekeeping operation and the type of honey you want to produce.

- Frames should be made of a non-toxic material, such as wood or plastic.

- Foundation should be made of a thin layer of wax or plastic.

- The smoker should be filled with dry, loose material, such as pine needles or straw.

- The hive tool should be made of a strong, durable material, such as stainless steel.

- Protective gear should be comfortable and fit well.

- Other supplies, such as a honey extractor, hive mat, and hive stand, should be made of durable materials and easy to use.

By following these tips, you can choose the right beekeeping equipment and use it safely and effectively.

4
BEE SAFE! BEE PREPARED! BEEREAL! APIARY LAWS & REGULATIONS

"A, Bee, C, D."
- *The Alphabet*

Beekeeping can be a rewarding hobby, but it is important to be aware of the risks involved. Bees can sting, and if you are not careful, you can be seriously injured.

Here are some tips for beekeeping safety:
- **Wear protective clothing**: When you are working with bees, always wear protective clothing, including a veil,

gloves, and long sleeves. This will help to protect you from bee stings.

  - The veil should cover your face and neck, and it should be made of a material that bees cannot easily penetrate.
  - The gloves should be made of a thick material that will protect your hands from bee stings.
  - The long sleeves should be made of a loose-fitting material that will not restrict your movement.

- **Bee calm**: Bees are less likely to sting if you are calm. If you start to panic, the bees will sense your fear and they may become more aggressive.
  - Take deep breaths and try to relax.
  - If you start to feel anxious, take a break and come back to the hive later.

- **Don't swat at bees**: If a bee stings you, don't swat at it. Swatting will only make the bee more likely to sting you again.
  - Instead, gently brush the bee away.

- **Remove the stinger**: If a bee stings you, remove the stinger as quickly as possible. The stinger contains venom, and if you leave it in, the venom will continue to be released.
  - Use a credit card or a piece of paper to scrape the stinger out.

- **Treat the sting**: If you are stung by a bee, treat the sting with an ice pack or cold compress. You can also apply a topical cream or ointment to help relieve the pain and swelling.
  - If you are allergic to bees, you may also need to take an antihistamine or other medication.

In addition to these general tips, there are a few other things you can do to stay safe when beekeeping:

- **Inspect your hives regularly**: Inspect your hives regularly to make sure that they are healthy and that there are no problems. This will help you to identify any potential problems early on and take steps to correct them.

   - Check for signs of pests or diseases.
   - Make sure that the hive is properly ventilated.
   - Check the level of honey in the hive.

- **Don't smoke the bees**: Smoking bees is a common practice, but it is not necessary and it can actually make the bees more aggressive. If you do smoke the bees, use it sparingly and only when necessary.
   - Smoke can be used to calm the bees, but it can also make them more aggressive.

- **Be aware of your surroundings**: When you are working with bees, be aware of your surroundings and make sure that you are not near any areas where people or pets might be. This will help to protect others from being stung by bees.
   - Keep children and pets away from the hive.

- If you are working near a road, make sure that you are visible to drivers.

By following these tips, you can help to reduce the risk of being stung by bees and stay safe while beekeeping.

Here is a story about a beekeeper who learned the importance of beekeeping safety the hard way:

John was a new beekeeper, and he was eager to get started. He had read all the books and watched all the videos, and he was confident that he knew what he was doing. John set up his hive in his backyard, and he was soon visited by a swarm of bees. The bees were happy to have a new home, and they quickly started to build their hive. John was excited to watch the bees work, but he was also a little bit nervous. He had heard stories about beekeepers getting stung, and he didn't want to be one of them. One day, John was inspecting his hive when he accidentally bumped into the hive. The bees were startled, and they started to swarm. John panicked and started to swat at the bees. The bees were not happy, and they stung John several times. John was in pain, and he had to run to the house to get help. John learned his lesson the hard way. He learned that it is important to be calm when working with bees, and he learned that it is important to wear protective clothing. John's story is a reminder that beekeeping can be a dangerous hobby if you are not careful. John died.

However, by following the tips in this chapter, you can help to reduce the risk of being stung by bees and stay safe while beekeeping.

Here are some additional tips for beekeeping safety:

- If you are allergic to bees, be extra careful. If you are stung by a bee and you have an allergic reaction, you could go into anaphylactic shock and die.

- Never work with bees alone. Always have someone with you in case you need help.

- If you are stung by a bee, seek medical attention immediately. Even if you are not allergic to bees, a sting can still be dangerous.

By following these tips, you can help to ensure your safety while beekeeping.

Beekeeping is a rewarding hobby, but it is important to be aware of the laws and regulations that govern beekeeping in your area. These laws and regulations are in place to protect the bees, the public, and the environment.

Most states require beekeepers to obtain a permit or license before they can keep bees. These permits or licenses typically require beekeepers to provide information about their beekeeping practices, such as the number of hives they keep and the location of their hives.

The requirements for obtaining a permit or license vary from state to state. In some states, you can simply apply for a permit online. In other states, you may need to attend a beekeeping class or pass a beekeeping exam.

Some states also have laws that restrict the location of beehives. For example, some states prohibit beehives from being located within a certain distance of schools or hospitals.

These restrictions are in place to protect the public from bee stings. Bees can be aggressive, and they can sting if they feel threatened. By restricting the location of beehives, these laws help to reduce the risk of bee stings.

Bees are susceptible to a number of diseases and pests. If a beehive is infected with a disease or pest, it can spread to other beehives and cause widespread damage.

To help prevent the spread of diseases and pests, some states require beekeepers to report any cases of disease or pest to the state's department of agriculture.

In addition to the laws and regulations mentioned above, there are a number of other laws and regulations that may apply to beekeeping in your area. For example, some states have laws that require beekeepers to use non-toxic pesticides.

It is important to be familiar with all of the laws and regulations that govern beekeeping in your area. You can find information about beekeeping laws and regulations on the websites of your state's department of agriculture or apiary inspection service.

Here are some tips for staying safe and compliant with beekeeping laws and regulations:

- **Do your research**: Before you start beekeeping, it is important to do your research and learn about the laws and regulations that govern beekeeping in your area.

- **Get a permit or license**: In most states, you will need to obtain a permit or license before you can keep bees.

- **Follow the rules**: Once you have a permit or license, it is important to follow the rules that govern beekeeping in your area.

- **Be prepared for emergencies/crises**: Beekeeping can be a dangerous hobby, so it is important to be prepared for emergencies. Make sure you have a plan in place in case you are stung by a bee, or if your hive swarms.

By following these tips, you can help to ensure that you are safe and compliant with beekeeping laws and regulations.

Here are some additional tips for staying safe and compliant with beekeeping laws and regulations:

- **Know your neighbors**: If you live in a densely populated area, it is important to introduce yourself to your neighbors and let them know that you are a beekeeper. This will help to reduce any concerns they may have about bees.

- Be mindful of your surroundings: When you are working with bees, be mindful of your surroundings and make sure that you are not near any areas where people or pets might be.

- Use proper safety equipment: When you are working with bees, it is important to use proper safety equipment, such as a veil, gloves, and long sleeves.

- Be patient: Beekeeping is a learning process, so be patient and don't get discouraged if you don't get it right away.

By following these tips, you can help to ensure that you are safe and compliant with beekeeping laws and regulations.

5
NOT JUST BEE CUM (HOW TO HARVEST HONEY)

"Would wanting to fuck a bee be kinda like beastiality? Does that apply to insects? Can somebody answer that?"

\- *Researchers*

Harvesting honey from a beehive is a rewarding experience, but it is important to do it safely and effectively. Here are some tips on how to harvest honey from a beehive.

The best time to harvest honey is in the fall, when the bees have finished collecting nectar for the season. However, you can also harvest honey in the spring, if the hive is producing a lot of honey.

The reason why fall is the best time to harvest honey is because the bees will have stored up a surplus of honey for the winter. This means that there will be plenty of honey for you to harvest without harming the bees.

Before you harvest honey, you need to prepare the hive. This includes:

- **Wearing protective gear**: You should always wear protective gear when working with bees, including a veil, gloves, and a bee suit. This will protect you from being stung by the bees.

- **Smoking the hive**: Smoking the hive will calm the bees and make them less likely to sting. This is done by burning some dry material, such as pine needles or straw, in a smoker. The smoke creates a cloud of smoke that masks the bees' alarm pheromones.

- **Removing the frames**: You will need to remove the frames from the hive to harvest the honey. This is done by using a hive tool to pry open the hive and then removing the frames.

Once you have removed the frames from the hive, you can harvest the honey. There are two main methods for harvesting honey:

- **Crush and strain**: This is the simplest method for harvesting honey. You simply crush the honeycomb and strain the honey out. This method is not as efficient as the extractor method, but it is a good option if you do not have an extractor.

- **Extractor**: An extractor is a machine that spins the frames, which causes the honey to be drawn out. This is a more efficient method for harvesting honey, but it requires a special machine.

If you are using an extractor, you will need to place the frames in the extractor and then turn on the machine. The honey will be drawn out of the frames and into a container.

Once you have harvested the honey, you need to store it properly. Honey can be stored in jars or plastic containers. It is important to store honey in a cool, dark place. This will help to preserve the flavor and quality of the honey.

If you are having trouble harvesting honey from your hive, there are a few things you can do:

- **Check the weather**: If it is too hot or too cold, the bees will not be producing honey.

- **Check the hive**: Make sure that the hive is in a sunny spot and that it is protected from the wind.

- **Check the bees**: If the bees are not healthy, they will not be able to produce honey.

Here are some additional tips for harvesting honey from a beehive:

- **Start with a few frames**: If you are new to beekeeping, it is best to start with a few frames of honey. This will help you to learn how to harvest honey without overwhelming the bees.

- **Be patient**: It takes time and patience to harvest honey from a beehive. Don't get discouraged if you don't get a lot of honey the first time. Just keep practicing and you will eventually get the hang of it.

- **Have fun**! Harvesting honey from a beehive is a fun and rewarding experience. Enjoy the process and savor the delicious honey that you produce.

6
HOW TO PREVENT & MANAGE COLONY COLLAPSE DISORDER

"Pollinator? More like HOLLAnator!"

- Gwen, unprompted

Colony Collapse Disorder (CCD) is a serious threat to honeybees and the pollination services they provide. In CCD, worker bees in a honey bee colony disappear, leaving behind a queen, plenty of food, and a few nurse bees to care for the remaining brood. The exact cause of CCD is still unknown, but it is thought to be a combination of factors, including pesticides, mites, and poor nutrition.

There are a number of things that can be done to help prevent CCD, including:

- **Using less pesticides**: Pesticides are a major cause of CCD, so it is important to use them sparingly. If you must use pesticides, choose those that are less harmful to bees.

Pesticides can kill bees directly, or they can weaken them so that they are more susceptible to disease. They can also disrupt the bees' ability to communicate and coordinate their activities.

For example, neonicotinoids are a class of pesticides that are known to be harmful to bees. These pesticides are often used on crops that are attractive to bees, such as corn and soybeans.

- **Managing mite populations**: Mites are another major cause of CCD. Mites are tiny parasites that feed on the blood of bees. They can weaken the bees and make them more susceptible to disease.

There are a number of ways to manage mite populations, including using miticides, hive sanitation, and breeding mite-resistant bees.

Miticides are chemicals that can be used to kill mites. However, miticides can also be harmful to bees, so they should be used only as a last resort.

Hive sanitation is the process of cleaning and disinfecting a hive. This helps to remove mites and other pests from the hive.

Breeding mite-resistant bees is a long-term solution to the problem of mite infestation. However, it takes time and effort to breed mite-resistant bees.

- **Providing adequate nutrition**: Bees need a balanced diet to stay healthy. They need a variety of pollen and nectar to get the nutrients they need. They also need water.

If bees do not have a balanced diet, they can become malnourished and more susceptible to disease.

Pollen is a good source of protein and carbohydrates for bees. Nectar is a good source of sugar for bees. Bees also need water to stay hydrated.

Beekeepers can provide bees with a balanced diet by planting a variety of flowers in their apiaries. They can also supplement the bees' diet with sugar water or pollen patties.

- **Managing stress**: Bees can be stressed by a number of factors, including overcrowding, poor weather, and transportation. It is important to manage these stressors to help keep your bees healthy.

Overcrowding can lead to competition for resources, which can stress the bees. Poor weather can make it difficult for the bees to forage for food and water. Transportation can also be stressful for bees, as it can disrupt their social order and expose them to new pathogens.

Beekeepers can help to manage stress in their bees by providing them with adequate space, by monitoring the weather, and by transporting the bees carefully.

If you do experience CCD, there are a number of things you can do to manage the situation, including:

- **Inspecting your hives regularly:** This will help you to identify any problems early on. If you see signs of CCD, such as a sudden loss of worker bees, you should inspect your hives regularly to see if there are any other problems.

- **Requeening**: If the queen is dead or unhealthy, you may need to requeen the hive. A new queen can help to revitalize the hive and get it back on track.

- **Supplementing the hive with food**: If the hive is running low on food, you may need to supplement it. You can do this by providing them with sugar water or pollen patties.

- **Moving the hive**: If the hive is in a stressful environment, you may need to move it to a more favorable location. For example, if the hive is located near a pesticide-intensive farm, you may need to move it to a more natural area.

- **Treating the hive with miticides**: If the hive is infested with mites, you may need to treat it with miticides. However, miticides should only be used as a last resort, as they can be harmful to bees.

- **Contacting a beekeeper society**: If you are not sure how to manage CCD, you can contact a beekeeper association for help.

7
THE FUTURE OF BEES

"bzzzz bbzbb zzzzbbb bbbzzzzzb ubzbzbuzzzzz"

- bee

Beekeeping is a centuries-old practice that is essential to our food system. Bees pollinate crops that provide us with food, medicine, and other products. However, bees are facing a number of challenges, including climate change, habitat loss, and pesticide use.

As a result, the future of beekeeping is uncertain. However, there are a number of things that can be done to help ensure the survival of bees and the future of beekeeping.

One important step is to educate the public about the importance of bees and the challenges they face. We need to raise awareness of the need to protect bees and their habitats. We can do this through school programs, public education campaigns, and social media.

We also need to develop more sustainable beekeeping practices. This includes using less pesticides, providing bees with better nutrition, and managing hives in a way that is more natural. We can do this by working with beekeepers to develop new methods and technologies.

Finally, we need to support research into bee health. This research will help us to better understand the challenges that bees face and develop new ways to help them survive. We can do this by funding research and providing grants to beekeepers and scientists.

The future of beekeeping is uncertain, but it is not bleak. With education, innovation, and support, we can ensure that bees continue to play a vital role in our food system.

<u>To reiterate</u> the beginning of this book, here are some of the challenges that bees are facing:

- **Climate change**: Climate change is causing temperatures to rise and weather patterns to become more extreme. This is making it difficult for bees to survive, as they need specific temperatures and conditions to thrive.

- **Habitat loss**: Bees need a variety of habitats to survive, including flowers, trees, and water. However, habitat loss is a major problem, as development and agriculture are destroying bee habitats.

- **Pesticide use**: Pesticides are a major threat to bees, as they can kill bees directly or make them more susceptible to disease.

- **Varroa mite**: Varroa mites are a parasitic mite that can kill bees. Varroa mites are a major problem for beekeepers, and they are becoming increasingly resistant to treatment.

Here are some of the things that can be done to help bees:

- **Plant bee-friendly plants**: Bees need flowers to survive, so planting bee-friendly plants in your garden or yard can help to support local bee populations. There are many different types of bee-friendly plants, so you can find some that will work well in your climate and landscape.

- **Reduce your use of pesticides**: Pesticides are a major threat to bees, so reducing your use of pesticides can help to protect bees. There are many non-toxic alternatives to pesticides, so you can find some that will work for your garden.

- **Support beekeepers**: Beekeepers are essential to the survival of bees, so supporting beekeepers can help to ensure that there are enough bees to pollinate our crops. You can support beekeepers by buying local honey, or by donating to beekeeping organizations.

- **Educate others about bees**: The more people who understand the importance of bees, the more likely we are to take action to protect them. You can educate others about bees by talking to your friends and family, or by volunteering at a local beekeeping organization.

If you are interested in beekeeping, I encourage you to learn more about it. There are many resources available online and in libraries. You can also find beekeeping clubs in your area that can provide you with support and advice.

Together, we can help to ensure the future of honey bees and the pollination services they provide. Thank you for reading!

The future of honey bees is in our hands. We can choose to ignore the problem and let the bees disappear, or we can choose to take action and help them survive.

I urge you to take action. Here are some things you can do:

- **Learn** about CCD and the threats to honey bees.

- **Support** beekeepers & encourage them to continue their important work.

- **Make changes** to your own life to reduce your impact on honey bees.

- **Educate** others on the importance of honey bees.

Together, we can make a difference. :)

This rest of this chapter is reserved to bring all of you fellow bee lovers an almost-complete list of higher education institutions that provide apiary & beekeeping programs. See below to bumble into your future education!

Schools

AUBURN UNIVERSITY

DEPARTMENT OF ENTOMOLOGY & PLANT PATHOLOGY

BRITISH COLUMBIA-SIMON FRASER UNIVERSITY

DEPARTMENT OF BIOLOGICAL SCIENCES

CLEMSON UNIVERSITY-SOUTH CAROLINA

BEE INFORMATION

DEPARTMENT OF ENTOMOLOGY, SOILS & PLANT SCIENCES

COLORADO STATE UNIVERSITY

DEPARTMENT OF BIOAGRICULTURAL SCIENCES & PEST MANAGEMENT

CORNELL UNIVERSITY

COLLEGE OF AGRICULTURE & LIFE SCIENCES

MASTER BEEKEEPER

FLORIDA A&M UNIVERSITY

ENTOMOLOGY & STRUCTURAL PEST CONTROL PROGRAM

IOWA STATE UNIVERSITY OF SCIENCE AND TECHNOLOGY

DEPARTMENT OF ENTOMOLOGY

KANSAS STATE UNIVERSITY

DEPARTMENT OF ENTOMOLOGY

LOUISIANA STATE UNIVERSITY

DEPARTMENT OF ENTOMOLOGY

MCGILL UNIVERSITY

DEPARTMENT OF NATURAL RESOURCE SCIENCES

MICHIGAN STATE UNIVERSITY

DEPARTMENT OF ENTOMOLOGY

MSU ENTOMOLOGY

MISSISSIPPI STATE UNIVERSITY

BEEKEEPING TIPS-DESCRIPTION: NOTES AND NEWS ON BEEKEEPING

CLARENCE COLLISON-ENTOMOLOGY & PLANT PATHOLOGY

DEPARTMENT OF ENTOMOLOGY & PLANT PATHOLOGY

UNIVERSITY OF MONTANA

MASTER BEEKEEPING CERTIFICATE

NEW MEXICO STATE UNIVERSITY

DEPARTMENT OF ENTOMOLOGY, PLANT PATHOLOGY & WEED SCIENCE

NORTH CAROLINA STATE UNIVERSITY

DEPARTMENT OF ENTOMOLOGY

NORTH DAKATO STATE UNIVERSITY

DEPARTMENT OF ENTOMOLOGY

OHIO STATE UNIVERSITY

BEE AND POLLINATION

DEPARTMENT OF ENTOMOLOGY

DR. JAMES E. TEW

HONEY BEE LABORATORY

OKLAHOMA STATe UNIVERSITY

DEPARTMENT OF ENTOMOLOGY & PLANT PATHOLOGY

OREGON STATE UNIVERSITY

DEPARTMENT OF ENTOMOLOGY

PENN STATE UNIVERSITY

DEPARTMENT OF ENTOMOLOGY

MID-ATLANTIC APICULTURE RESEARCH AND EXTENSION CONSORTIUM

PURDUE UNIVERSITY

DEPARTMENT OF ENTOMOLOGY

RUTGERS UNIVERSITY, COOK COLLEGE

DEPARTMENT OF ENTOMOLOGY

MID-ATLANTIC APICULTURE RESEARCH AND EXTENSION CONSORTIUM

SOUTH DAKOTA STATE UNIVERSITY

DEPARTMENT OF PLANT SCIENCE

TEXAS A&M UNIVERSITY

DEPARTMENT OF ENTOMOLOGY

TEXAS TECH UNIVERSITY

DEPARTMENT OF PLANT & SOIL SCIENCE

THE UNIVERSITY OF ARIZONA

DEPARTMENT OF ENTOMOLOGY

THE UNIVERSITY OF GEORGIA

DEPARTMENT OF ENTOMOLOGY

University of Georgia Honey Bee Program

THE UNIVERSITY OF TENNESSEE

DEPARTMENT OF ENTOMOLOGY & PLANT PATHOLOGY

UNIVERSITY OF ALBERTA

DEPARTMENT OF BIOLOGICAL SCIENCES

UNIVERSITY OF ARKANSAS

DEPARTMENT OF ENTOMOLOGY

UNIVERSITY OF CALIFORNIA-BERKLEY

DIVISION OF INSECT BIOLOGY

UNIVERSITY OF CALIFORNIA-DAVIS

BEE BRIEFS

DEPARTMENT OF ENTOMOLOGY

DR. ERIC MUSSEN

DR. ERIC MUSSEN'S NEWSLETTERS

THE AMERICAN ASSOCIATION OF PROFESSIONAL APICULTURISTS

UNIVERSITY OF CALIFORNIA-RIVERSIDE

DEPARTMENT OF ENTOMOLOGY

UNIVERSITY OF CONNECTICUT

DEPARTMENT OF PLANT SCIENCES

UNIVERSITY OF DELAWARE

DEPARTMENT OF ENTOMOLOGY & WILDLIFE ECOLOGY

DR. DEBORAH A DELANEY

MID-ATLANTIC APICULTURE RESEARCH AND EXTENSION CONSORTIUM

UNIVERSITY OF FLORIDA

DEPARTMENT OF ENTOMOLOGY & NEMATOLOGY

UNIVERSITY OF GUELPH

DEPARTMENT OF ENVIRONMENTAL BIOLOGY

UNIVERSITY OF HAWAII AT MANOA

DEPARTMENT OF PLANT & ENVIRONMENTAL PROTECTION SCIENCES

UNIVERSITY OF IDAHO

DEPARTMENT OF PLANT, SOIL & ENTOMOLOGY SCIENCES

UNIVERSITY OF ILLINOIS AT URBANA-CHAMPAIGN

DEPARTMENT OF ENTOMOLOGY

UNIVERSITY OF KENTUCKY

DEPARTMENT OF ENTOMOLOGY

UNIVERSITY OF MAINE

DEPARTMENT OF BIOLOGICAL SCIENCES

UNIVERSITY OF MANITOBA

DEPARTMENT OF ENTOMOLOGY

UNIVERSITY OF MARYLAND

DEPARTMENT OF ENTOMOLOGY

MID-ATLANTIC APICULTURE RESEARCH AND EXTENSION CONSORTIUM

UNIVERSITY OF MASSACHUSETTS

DEPARTMENT OF ENTOMOLOGY

UNIVERSITY OF MINNESOTA

DEPARTMENT OF ENTOMOLOGY

UNIVERSITY OF MISSOURI

DEPARTMENT OF ENTOMOLOGY

UNIVERSITY OF NEBRASKA

DEPARTMENT OF ENTOMOLOGY

UNIVERSITY OF RHODE ISLAND

DEPARTMENT OF PLANT SCIENCES

UNIVERSITY OF VERMONT

DEPARTMENT OF ENTOMOLOGY

UNIVERSITY OF WISCONSIN MADISON

DEPARTMENT OF ENTOMOLOGY

UNIVERSITY OF WYOMING

DEPARTMENT OF PLANT SCIENCES

UTAH STATE UNIVERSITY

DEPARTMENT OF BIOLOGY

VIRGINIA TECH

DEPARTMENT OF ENTOMOLOGY

THE SMALL HIVE BEETLE-DESCRIPTION: ALL ABOUT THE NEW HONEYBEE PEST

WASHINGTON STATE UNIVERSITY

APIS.WSU.EDU

DEPARTMENT OF ENTOMOLOGY

DR. STEVE SHEPPARD

WEST VIRGINIA UNIVERSITY

DEPARTMENT OF PLANT & SOIL SCIENCE

MID-ATLANTIC APICULTURE RESEARCH AND EXTENSION CONSORTIUM

PROTECTING HONEYBEES FROM PESTICIDE POISONING

BONUS POETRY APPENDIX

Bees

They flit from flower to flower,

A buzz of wings and hum,

A murmur in the air,

A sweet and gentle hum.

They gather pollen and nectar,

To make their honey sweet,

A gift from nature's hand,

A taste of summer heat.

They work together in harmony,

A hive of industry,

A buzzing, humming mass,

A sight to see and hear.

They are the pollinators,

Of our flowers and crops,

They help to keep the world alive,

With their hard work and toil.

So let us celebrate the bees,

These tiny creatures,

Who do so much for us,

And give us so much pleasure.

A Sonnet for Bees

Oh, bees, so small but mighty,

With your work so very sweet,

You pollinate the flowers,

And make the world complete.

You are a symbol of industry,

And your work is never done,

You toil from dawn to dusk,

To keep the world in tune.

You are a vital part of nature,

And we are so grateful for you,

So thank you, bees, for all you do,

We wouldn't be here without you.

A Plea for Bees

Oh, bees, please stay,

We need you so,

To pollinate our flowers,

And keep our world alive.

We've done so much to harm you,

With pesticides and herbicides,

But we must stop,

Before it's too late.

We need you, bees,

To help us survive,

So please stay,

And help us save our world.

ABOUT THE AUTHORS

Author Connor Gleim in Oregon conducting Apiary research

Connor Gleim received his Doctor of Philoso-Bee Beegree from the Master Beekeeping Program at Oregon State University. While completing his degree, Gleim worked with the United Nations to establish a more concrete Apiary Law structure in America. His research on the United Nations and their bee-saving efforts is set for release in 2024. In the meantime, he will be buzzing around with his wife – Dep – also a Beekeeper. They reside in Oregon.

Author Austin Robinson pointing at a bunch of bees

Austin Robinson received a Brazilian Butt Lift from the pre-med students at Portland State University in Portland, Oregon. He has numerous apiary accomplishments fighting justice in the Beekeeping Disputes Court. His favorite activities include educating children at elementary schools about bees, doing pro bono work for poor apiary researchers, and – of course – beekeeping in Oregon!

Author Patrick Golden in front of a Blockbuster

Patrick Golden received his diagnosis of Attention Deficit Hyperactivity Disorder after serving 20 years in the United States Marines (unrelated). While stationed on the seaside of many countries, Golden discovered his love of having a mental illness when he started not knowing how to pay attention. He currently travels with 'People With ADHD Without Borders' across North Africa, the Arabian Peninsula, and the islands surrounding Southeast Asia. He has inspired thousands of impoverished children who he thinks maybe has ADHD!

Made in the USA
Monee, IL
04 November 2026

67375807R00049